Generis
PUBLISHING

AF414288

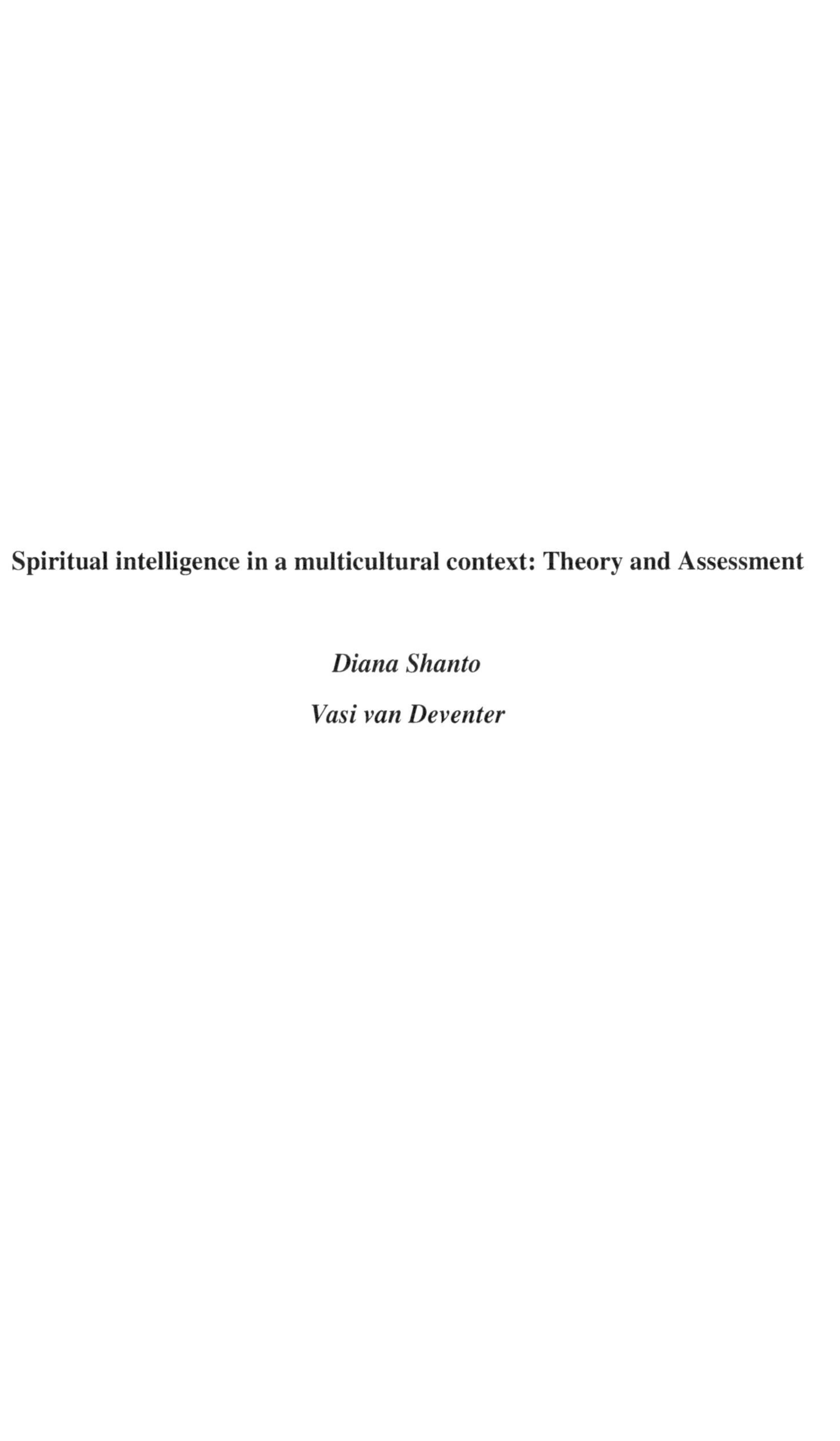

Spiritual intelligence in a multicultural context: Theory and Assessment

Diana Shanto

Vasi van Deventer

CIP a Camerei Naţionale a Cărţii

Diana Shanto, Vasi van Deventer

Spiritual intelligence in a multicultural context: Theory and Assessment / Diana Shanto, Vasi van Deventer. – Chişinău : Generis Publishing (Online Marketing Group), 2020 (Print on demand). – 57 p. : fig., tab.
Referinţe bibliogr.: p. 41-48.

ISBN 978-9975-3429-7-1.

316.344(661.2) S 52

Cover image: www.pixabay.com

Generis Publishing
Online orders: www.generis-publishing.com
Orders by email: info@generis-publishing.com

CONTENTS

List of tables

Figures

Abstract

This study investigates the nature of spiritual intelligence and its link to ethnic identity, and gauges the difference across the main ethnic groups in Mauritius: Hindu-Mauritians, Creole-Mauritians, and Muslim-Mauritians. The authors proposed a new scale, the Multicultural Spiritual Intelligence Scale (MSIS), using the following six dimensions: self-awareness, transcendental awareness, levels of consciousness, quest for meaning, sensitivity, and resilience. The MSIS's construct validity was tested with other similar scales. The Balanced Inventory of Desirable Responding (BIDR) (Paulhus, 1991) was also included to check for socially desirable responses and MSIS's divergent validity. A Welch ANOVA revealed a statistical difference in spiritual intelligence among the ethnic groups: Welch's F (2, 639.98) = 3.923. Spearman's rank order correlation revealed that ethnic identification was connected to spiritual intelligence: r_s (98) = 0.52, $p < 0.0005$. A Games-Howell post-hoc analysis indicated a statistically significant mean difference between Muslim-Mauritians and Hindu Mauritians (0.27, 95% CI [0.083, 0.45]) and between Muslim-Mauritians and Creole Mauritians (0.44, 95% CI [0.25, 0.62]). The Muslim-Mauritians obtained the highest score in both ethnic exploration and ethnic commitment. Implications for further research include using the MSIS with other minority groups in Mauritius.

Keywords: spiritual intelligence, multicultural, ethnic identity

Dedication

To Anand and Pat for their love and support

Acknowledgements

This book is about spiritual intelligence and I want to thank all the people in my life – parents, siblings, friends and colleagues- all those who have supported me in some way.

A massive thank you to Prof Vasi van Deventer for his invaluable support and guidance.

Introduction

The concept of spiritual intelligence sprouts from the theory of multiple intelligences, and emphasises the specific criteria which have to be fulfilled for an aspect of intelligence to be defined as spiritual intelligence. Eight main criteria were ugge ed for n in elligence o be defined   uch (Arm rong, 2009; G rdner, 2006, 2011). The fir   condi ion w   prob ble i ol ion by br in d m ge;   econd, the presence of savants, prodigies, and other outstanding individu l ; hird,   ypic l development history and a definable set of expert 'end-state' (Armstrong, 2009, p. 149) func ioning; four h, n evolu ion ry hi ory nd evolu ion ry credibili y; fif h, v lid   ion from p ychome ric ou come ; ix h,   u hen ication from experimental p ychologic l   k ;   even h, recogni ble core oper ion( );   nd fin lly, predisposition to encoding in a symbol system (Gardner, 1999, 2011). Spiritual intelligence (SQ) is the latest addition to the list of intelligences. King and DeCicco (2009) defined SQ as a mental ability that contributes to spiritual development, and observed that it meets the criteria for intelligence. Firstly, SQ involves a set of mental capacities that differ from preferred behaviours. Secondly, SQ develops throughout the lifespan. Thirdly, the adaptive function of spirituality has been evidenced in several studies. An analysis of its neurological foundation revealed that SQ unified all other intelligences, such as EQ[1] and IQ (Zohar & Marshall, 2001).

Gardner (2006) proposed enlarging the scope of human potential beyond the limits of the IQ score. He argued that there were three biases that restraint the perspective on the wide-ranging human abilities. The Western values dominance, the overreliance on abilities that were testable and overemphasis on one intelligence such as the logical-mathematical ability. He also posited intelligence as a 'biopsychological potential to process information' (1999, pp. 33–34) which can be triggered in a cultural background to solve problems or generate products which are of value in a culture. Gardner (2006) introduced the concept of multiple intelligences (MI), which viewed intelligence as a 'computational capacity' (p. 5) stemming from human biology and psychology. MI theory, in contrast to the classical view, described intelligence as the universal human capacity to process information and the ability to solve problems involving both anticipation and goal setting (Gardner, 2006).

Many studies on SQ associate it with organisational variables such as leadership and job satisfaction, but few look at ethnic profiles. Ethnicity is part of Mauritius, a country which strives to promote national identity over and above all.

---

[1] EQ stands for emotional quotient while IQ refers to the intellectual quotient.

In a multi-ethnic society, ethnic identity is a relevant social identity (Greig, 2003; Umaña-Taylor, 2011, as cited in Umaña-Taylor, 2015), and essential to the mental functioning of ethnic and racial minority groups' members (Phinney, 1990). Deep ethnic identification with one's ethnic background has been steadily associated with a multitude of positive outcomes such as meaning in life, lower incidence of depression, and 'externalizing behaviours' (Umaña-Taylor et al., 2014, p. 170; see also Tummala-Narra, 2015), superior academic achievement, social competence, and better management of discriminatory conditions (Juang & Syed, 2008; Martinez & Dukes, 1997; Mossakowski, 2003; Umaña-Taylor et al. 2014). There is a dearth of studies on the correlation between ethnic identification and spiritual intelligence.

Ethnicity is a multidimensional social boundary established by people's consciousness of genuine or alleged shared pedigree (Schermerhorn, 1970, as cited in Calvillo & Bailey, 2015), cultural practices, and the distinctness of a community (García Coll, Crnic, Lamberty, Wasik, Jenkins, Vázquez García & McAdoo, 1996; Helms, 2007). Umaña-Taylor (2015) posited that a person has multiple social identities, which include both cognitive and affective components. Ethnic identity denotes allegiance to a cultural group and dedication to its cultural practices (Helms, 2007). In this book, ethnicity is not conceived as a biologically determined phenomenon but as a social construction generated by individuals as they affiliate themselves with specific ethnic groups (McKinlay & McVittie, 2011). The ethnic identity developmental paradigm means a convergence of Eriksonian/Marcian identity theory and social identity theory (Pahl & Way, 2006; Phinney, 1990; Syed & Juang, 2014). Phinney (1990) described Tajfel and Turner's (1979) social identity theory as advancing the idea that group membership bestows a sense of belongingness that promotes a positive self-concept. Phinney and Ong's multidimensional developmental paradigm of ethnic identity underscored the fundamental processes of exploration and commitment which were fashioned over time (Phinney & Ong, 2007; Syed, Walker, Lee, Umaña-Taylor, Zamboanga, Schwartz, Armenta & Huynh, 2013). Ethnic identity exploration concerns the process of searching for, examining, and gaining knowledge and experiences about the meaning of one's ethnic origin. Ethnic identity commitment is the process of developing a sense of attachment or belongingness (Phinney, 1990) and an affective bond to one's ethnic group (Pahl & Way, 2006; Phinney & Ong, 2007; Syed et al., 2013).

Mauritius is part of the Mascareignes Archipelago located 890 kilometers to the east of Madagascar (Government Portal, 2015) in the South West Indian Ocean (Jungers, Gregoire & Slagel, 2009). The island of Mauritius has a 'cosmopolitan culture' (Government portal, 2015) and it is usually dubbed the 'Mauritian rainbow' (Jungers & Gregoire, 2010, p. 84) for its rich diversity of races, ethnicities, languages, and religions. The Constitution of Mauritius acknowledges four main ethnic categories: Hindus, who represent 52 percent of the population; Muslims, with 16 percent; Sino-Mauritians, who represent only 3 percent; and the general population, which constitutes 29 percent (Carpooran, 2003; Ng Tseung Wong & Verkuyten, 2015). The general population, a 'catchall category' (Jungers et al., 2009, p. 302) has a negative connotation and includes any person who does not identify with the three other ethnic groups (Carpooran, 2003), such as the Creoles and the Franco-Mauritians (Addison & Hazareesingh, 1999). The general population is devoid of any religious legacy and is not related to ancestral countries like the other groups (Jungers, et al., 2009). The Muslim population was initially counted with the Hindus under one category—the Indo-Mauritians. The separation based on religious identity propelled its recognition as an ethnic group on its own in the Mauritian Constitution. Mauritius has an even wider cultural tapestry, considering the Hindu community is subdivided into Hindu, Tamil, Marathi, and Telegu people (Ng Tseung Wong & Verkuyten, 2015), who have their own temples, languages, dress code,[2] and food. The dominant group is the Hindu population, who came from the Indian continent to work as indentured labourers under British colonial rule (Addison & Hazareesingh, 1999). Most Muslims in Mauritius are Sunni, though a small percentage (approximately 5 percent) are Shi'a (Burrun, 2002). The Muslim community is not a unified population, but is divided across several lines. The recognition of ancestral languages and religious philosophy are the two primary issues on which the Muslim community diverges. There is a fine line between religions and ethnicity in Mauritius. The boundaries between the two are fuzzy.

Theory of spiritual intelligence

A scientific and universal definition of spiritual intelligence is proposed across six dimensions: self-awareness, transcendental awareness, level of consciousness, quest for meaning, sensitivity, and resilience. These dimensions have emerged in several studies. The qualities highlighted by Zohar and Marshall (2005) for the SQ development were instrumental in the scale development of this research. Among others, there were the self-awareness qualities, the search for meaning (quest for meaning), compassion (sensitivity) and the positive use of

---

[2] Indian clothing such as saris are worn differently by different ethnic groups.

adversity (resilience). King (2008) also contributed in his proposition of transcendental awareness and conscious state expansion (levels of consciousness) in the definition of SQ in this study (See also Amram & Dryer, 2008). Wigglesworth's (2006) suggestion of self-awareness as a major dimension with several stages of SQ development added to this research. Nasel (2004) and King (2008) also viewed SQ as partly a capacity for meaning production. The element of quest and practical aspects of spiritual intelligence were developed by Vaughan (2002). Wolman (2001) highlighted the existential capacity of the individual to seek meaning. The content of the MSIS-1 reflects the influence of Eastern, African and Western perspectives on the local population.

Self-awareness is a multifaceted notion with four general levels of organisation, specifically biogenetic, personal, social, and cultural-linguistic (Mascolo & Fischer, n.d., as cited in Ferrari, 1998). A high degree of self-awareness is a sign of spiritual intelligence (Zohar & Marshall, 2001). Self-awareness is accomplished in a relational context that is manifested as a more profound relationship with the divine Spirit and with other humans (DeHoff, 1998). Self-awareness is inclusive of mindfulness: it is a mental state marked by intense awareness (Holas & Jankowski, 2013) of one's thoughts, actions, or stimuli (Appel & Appel, 2009). Vaughan (2002) clearly stated that self-awareness is critical for the unleashing of spiritual maturity.

Sensitivity is the ability to feel empathy for others, to feel a sense of reverence for all living things and to recognise and admit one's mistakes. In Hay and Nye's (2006) model of children's spirituality three types of sensitivities were suggested, namely: awareness-sensing, mystery-sensing, and value-sensing. Awareness-sensing relates to the "here-and-now" (p.66) experience. It also involves "focusing" (p.70), which is defined as a respect for the bodily feelings. The notion of focused attention that gives way to a liberated feeling of accomplishment. Mystery sensing relates to the awareness of the incomprehensible. In mystery sensing, there is the experience of wonder, and awe. Value-sensing is lived as emotion such as delight and despair. Tirri and Nokelainen (2011) developed a spiritual sensitivity scale based on Hay and Nye's definition. However, the pool of items was too limited, with only 20 items, prior to EFA.

The third dimension denotes the capacity to perceive transcendent aspects of the self, of others, and of the physical world during the normal waking state of consciousness (King & DeCicco, 2009). Transcendence denotes an appreciation of or link to something beyond the self, which may or may not involve God or a transcendent attribute, but must include the characteristics of life which are filled with divine-like qualities such as 'immanence, boundlessness and ultimacy' (Pargament, 2013, as cited in King, Clardy & Ramos, 2014, p. 188). SQ implies an awareness of the transcendent (Vaughan, 2002). The interrogation 'Why?' (p. 4), according to Ramadan (2012), sets off the quest for meaning and a consciousness of our needs, limits, and power.

Ramadan (2012) observed that possible answers to existential questions were available either within the family or society or through an inward journey. The tendency to search for meaning is a characteristic of the spiritual human being (Zohar & Marshall, 2001). According to Zohar and Marshall (2001), anthropologists and neurobiologists claim that the desire for meaning forms symbolic thoughts, initiates language development, and results in the brain's growth. Batson and Schoenrade (1991) proposed a dimension in religion which they called "religion as quest" (p.417). The quest meant facing existential queries in their complexity. Spiritual growth was marked by the questioning of the established beliefs. Young, Cashwell and Woolington's (1998) study found a positive link between spirituality and moral development and meaning in life. According to Greenway, Phelan, Turnbull and Milne (2007), the act of questing gives the person a perspective for a more challenging spiritual experience and confronts existential questions without diminishing their complexity. The authors posited that questing individuals tend to view self-criticism and religious doubt as constructive experiences.

Searle (1998) defines consciousness as 'subjective states of awareness' (p. 4) which start when one wakes from a dreamless sleep and carry on until one falls asleep, goes into a coma, dies, or otherwise becomes unconscious. Tart (1975, as cited in King, 2008) differentiated between transcendental awareness and level of consciousness, asserting that transcendental awareness happens in normal waking state while levels of consciousness involve entering higher or spiritual states. Emmons (2000) concurred that moving in different states of consciousness is an ability to be recognised as a characteristic of SQ. The 'peak experiences' (p. 510) described by Maslow (1964, as cited in Feist & Feist, 2002) involve the movement in consciousness such that people tend to feel a 'disorientation in time and space, a loss of self-consciousness, an unselfish attitude' (p. 511) and have the possibility to transcend daily divergences.

Resilience originates from the Latin word *resilire*, which means 'to spring back' (Davoudi, 2013, p. 4). The American Psychological Association (2017) defined resilience as the process of adjusting well or 'bouncing back' in the face of hardship, trauma, disaster, threat, or even important causes of stress. Atkinson, Martin, and Rankin (2009) stated that resilience is the ability to get over the experiences of distress, deprivation, and danger. Resilience emanates from within the human soul or collective unconscious of the person, and from the external ecological and spiritual bases of strength (Richardson, 2002). The concept of the positive use of adversity advanced by Zohar and Marshall (2001, 2005) as an aspect of SQ is termed in this project as resilience. It is 'the capacity to withstand adversity or resist liability' (Kasen, Wickramaratne, Gameroff & Weissman, 2012, p. 509), as well as the potential possessed by individuals to prosper even after facing harsh situations (Dillen, 2012). However, Theron et al. (2013) observed that resilience cannot be conceptualised in a standardised way, as it is culturally specific. Clinton's (2008) paper on children who underwent traumatic experiences found that the internal features of resilience were hope, humour, a meaning in life, and a sense of connectedness. He defined resilience as both involving the successful use of coping strategies and to being transformed in a positive way.

Method: Online survey and paper-pencil survey

The objectives of this research were to first investigate the factor structure of spiritual intelligence. Secondly, it was important to determine whether the three ethnic groups differ with regard to spiritual intelligence. Thirdly, the study sought to establish how ethnic identity relates to spiritual intelligence. The level of ethnic identification among the three groups were also investigated. Finally, other variables such as religious groups and gender were observed in relation to SQ.

The five main research questions that guided the study were:

1 Do Hindu-Mauritians, Muslim-Mauritians, and Creole-Mauritians differ with regard to spiritual intelligence?
2 Is there a relationship between level of ethnic identity and spiritual intelligence?
3 Are there differences in ethnic identity in the three ethnic groups?
4 Do religious groups differ with regard to spiritual intelligence?
5 Are the differences in ethnic identity and spiritual intelligence explained by gender?

Participants

The sample population for the present study 1 were adults from the Republic of Mauritius. The target population or sampling frame for this study was working men and women above 18 who were in one of the three main ethnic groups: Mauritian Creoles, Mauritian Hindus, and Mauritian Muslims. The sample size was 1,177 respondents (47.7 percent male and 52.3 percent female), including participants who withdrew or submitted incomplete questionnaires. The sample population was initially set at 1,500 participants drawn from the working population. Appendix A describes the sample distribution. However, many questionnaires were not returned or not completed by participants. The three ethnic groups were fairly equally represented, with 330 Hindu-Mauritian participants (28 percent of the sample population), 308 Muslim-Mauritian participants (26.1 percent of the sample population), and 338 Creole-Mauritian participants (28.7 percent of the sample population). The participants who reported mixed, Franco-Mauritian, or Sino-Mauritian ethnicity represented 17.1 percent.

Measures:

MSIS-1, MEIM-R, MLQ, CD-RISC, PSC, NIRO and BIDR-6

The survey design used questionnaires, namely the Multicultural Spiritual Intelligence Scale (MSIS-1), the Multigroup Ethnic Identity Measure-Revised (MEIM-R), the Meaning in Life Questionnaire (MLQ), the Connor-Davidson Resilience Scale (CD-RISC), the Private Self-Consciousness Subscale (PSC), the New Indices of Religious Orientation—the extrinsic subscale (NIRO), and the Balanced Inventory of Desirable Responding (BIDR-6). The MSIS-1 (106 items), developed for this study, covers six dimensions: self-awareness (16 items), level of consciousness (19 items), transcendental awareness (21 items), sensitivity (16 items), resilience (16 items), and quest for meaning (18 items). The questionnaire also included a ten-item demographic section which inquired about the characteristics of the participants (Shaughnessy, Zechmeister, & Zechmeister, 2003), specifically gender, age, marital status, educational level, occupational background, religion, and religious affiliation. The items in the MSIS-1 were determined after a thorough review of the literature on spiritual intelligence scale development. The MEIM-R developed by Phinney and Ong (2007) has six items on ethnic identity and equally measured two dimensions of ethnic identity, namely exploration and commitment. The MSIS were compared to similar scales to investigate its construct validity. The selected measures for convergent validity were the Meaning in Life Questionnaire (MLQ) a 10-item self-reported measure with two subscales- the presence of meaning ($\alpha = 0.82$) and the search for meaning ($\alpha = 0.87$) (Steger, Frazier, Oishi & Kaler, 2006); the CD-RISC a 10-item self-report test with $\alpha = 0.9$ (Connor & Davidson, 2011); the Private Self-Consciousness Subscale (PSC) a 9-item measure with $\alpha = 0.75$ (Scheir & Carver, 1985); and the New Indices of Religious Orientation—the extrinsic subscale (NIRO) a 6-item measure ($\alpha = 0.97$) (Francis, 2007). The Balanced Inventory of Desirable Responding (BIDR-6) (Paulhus, 1991), a 40-item questionnaire with $\alpha = 0.83$, measures two constructs: self-deception enhancement (SDE) and impression management (IM). Each subscale of the BIDR has 20 items provided for the discriminant validity. Permission was sought from the respective authors to use these questionnaires. As Li and Bagger (2007) highlighted, it was not the test that is reliable but the test scores. In this research, the reliability coefficient for the MLQ scores was 0.845 (the standardised alpha value was 0.844), and the subscales (the presence of meaning and the search for meaning) had Cronbach's alpha values of 0.905 and 0.830 respectively. The CD-RISC and the PSC had also high internal consistency reliability scores of 0.924 and 0.785 respectively. The NIRO measure had a lower Cronbach's alpha value of 0.663 (the standardised alpha value was 0.662). The subscales of the NIRO such as compartmentalisation, social support, and personal support also had low reliability scores of 0.670, 0.542, and 0.676 respectively. The reliability coefficient of the BIDR scores was 0.770 for this study. The SDE and the IM subscales had reliability coefficients of 0.659 and 0.629 respectively.

Procedures: Survey

The questionnaires were mailed or administered online; this combination resulted in a high response rate (Creswell, 2012). The paper version was returned by 636 participants, who represented 54 percent of the sample. The companies in the study were randomly selected by drawing lots. For example, private clinics, private bus transport companies, state secondary schools and universities were part of the survey. While mailing the questionnaires was an expedient way of reaching a geographically scattered sample and resulted in relatively fast data collection, the researcher avoided the problem of few questionnaires being returned by collecting them personally. To achieve better reach for the survey, an online version of the questionnaire was also made available using eSurvey Creator (2015). The online version was available for a period of 12 months, with the option of filling it out in one go or over several intervals. This version garnered 541 participants (46 percent of the sample). Of those participants who completed the online version, 84 percent were invited by email, while only 16 percent (88 participants) responded to the page directly. Permission was sought to send email notifications to companies which provided the facility to its employees. The main drawback of the online and take-home administration was a weaker control over administrative procedures and from unauthorised use and distribution.

Results of study 1: Exploratory factor analysis of the spiritual intelligence scale

Cronbach's alpha coefficients were used as indices of the reliability of the questionnaire and its subscales. The Cronbach's alpha coefficient for the 106-item set was very strong, suggesting excellent internal consistency and reliability for the whole item pool (as shown in Table 1). The inter-item correlation matrix displayed a correlation coefficient of minimum 0.30, which was within the range of 0.15 and 0.50 for unidimensionality suggested by Clark and Watson (1995).

Table 1. Reliability of the MSIS-1 and its subscales (N = 1177)

Scales	No. of items	Cronbach alpha coefficient	Standardized alpha
MSIS	106	0.99	0.99
Self-awareness subscale	16	0.899	0.898
Level of consciousness subscale	19	0.923	0.923
Transcendental awareness subscale	21	0.946	0.946
Sensitivity subscale	16	0.916	0.916
Resilience subscale	16	0.934	0.933
Quest for meaning subscale	18	0.939	0.938

Table 2. Correlations among related and unrelated measures with MSIS-1 and its subscales

Measure: Variable	MSIS	SA	LC	TA	S	R	QM
MLQ (N= 1043)	0.679**	0.635**	0.608**	0.640**	0.631**	0.631**	0.678**
MLQ: Presence	0.722**	0.690**	0.637**	0.664**	0.680**	0.684**	0.715**
MLQ: Search	0.330**	0.291**	0.306**	0.331**	0.297**	0.292**	0.337**
CD-RISC (N= 1036)	0.828**	0.769**	0.784**	0.764**	0.772**	0.805**	0.776**
PSC (N= 1033)	0.699**	0.628**	0.654**	0.632**	0.685**	0.683**	0.664**
NIRO (N= 1031)	0.538**	0.509**	0.479**	0.527**	0.486**	0.531**	0.501**
NIRO: Compartmentalization	0.472**	0.458**	0.406**	0.456**	0.417**	0.487**	0.444*
NIRO: Social support	0.288**	0.257**	0.291**	0.290**	0.264**	0.264**	0.253**
NIRO: Personal support	0.389**	0.372**	0.325**	0.380**	0.360**	0.383**	0.375**
BIDR (N= 1024)	0.072*	0.082**	0.080*	0.093**	0.070*	0.043	0.038
BIDR: SDE	0.110**	0.121**	0.111*	0.123**	0.102**	0.086**	0.082**
BIDR: IM	0.019	0.025	0.032	0.044	0.024	-0.008	-0.012

Correlation was significant at the 0.01 level (2-tailed). * Correlation was significant at the 0.05 level (2-tailed). **MSIS = Multicultural Spiritual Intelligence Scale; **SA** = Self-awareness; **LC** = Level of Consciousness; **TA** = Transcendental awareness; **S** = Sensitivity; **R** = Resilience; **QM** = Quest for Meaning; **MLQ** = Meaning in Life; **CD-RISC** = Connor-Davidson Resilience Scale; **PSC** = Private self-consciousness subscale; **NIRO** = New Indices of Religious Orientation; **BIDR** = Balanced Inventory for Desirable Responding.

There was a strong correlation between the resilience dimension of the MSIS-1 and the CD-RISC, the Pearson's product-moment correlation coefficient was r (98) = 0.805, p < 0.0005 (see table 2.). The Pearson's product-moment correlation between the MSIS-1 and the BIDR was weak: r (90) = 0.072, p = 0.021. All the subscales of the MSIS-1 had low correlation with the BIDR and its subscales.

Exploratory factor analysis: Principal Component Analysis

A Principal Component Analysis (PCA) was performed on the 106-item questionnaire, which measured spiritual intelligence. The KMO measure, an index of sampling adequacy, was 0.988, which was considered 'marvellous' (Kaiser, 1974, as cited by Lund Research, 2013). Bartlett's test was statistically significant ($p < 0.0005$), which meant that the data was likely factorisable (Lund Research, 2013). A factor loading greater than 3.5 was considered salient in this study. The PCA unveiled 12 components that had eigenvalues greater than one, which accounted for 60.3 percent of the total variance. A varimax orthogonal rotation was used to facilitate interpretability. An inspection at the rotated component matrix revealed a 'complex structure' (Lund Research, 2013), which occurred when more than one component loaded on one variable. It was decided that all variables where more than one component loaded robustly were to be eliminated from the data. The interpretability criterion was also crucial for component retention. The interpretation of the data was fairly consistent with the initial proposition. The MSIS questionnaire was intended to assess six components. However, five components with moderate loadings met the interpretability criterion. The five-component solution described 51.4 percent of the total variance. Component loadings and communalities of the rotated solution are exhibited in Table 3. Comrey and Lee (1992) and Rummel (1970, as cited in Meyers, Gamst & Guarino, 2006) recommended that the labelling process of the factors be steered by those variables with high loadings (values of more than 0.65). Factor 1 was labelled resilience, factor 2 was renamed the existential quest, factor 3 was termed self-mastery, factor 5 was labelled transcendental awareness, and factor 6 was termed spiritual sensitivity.

Table 3. Rotated structure matrix for PCA with varimax rotation (major loadings for the MSIS > 0.35) (N = 1177)

Rotated Component Coefficients

Items	Factor 1	Factor 2	Factor 3	Factor 5	Factor 6	Communalities
Qu55	**.643**	0.204	0.210	0.161	0.183	0.635
Qu56	**.653**	0.216	0.226	0.147	0.105	0.674
Qu57	**.659**	0.211	0.116	0.044	0.145	0.591
Qu86	**.655**	0.290	0.169	0.085	0.158	0.664
Qu58	0.275	**0.420**	0.218	0.095	0.283	0.605
Qu70	0.262	**0.499**	0.252	0.247	0.143	0.590
Qu76	0.297	**0.402**	0.291	0.067	0.120	0.524

Qu83	0.235	**0.376**	0.081	0.251	0.250	0.540
Qu93	0.159	**0.622**	0.166	0.249	0.198	0.614
Qu97	0.277	**0.601**	0.242	0.188	0.156	0.615
Qu99	0.198	**0.646**	0.211	0.199	0.129	0.663
Qu101	0.271	**0.587**	0.205	0.173	0.188	0.636
Qu102	0.234	**0.622**	0.272	0.158	0.172	0.677
Qu105	0.084	**0.582**	0.249	0.184	0.147	0.545
Qu4	0.180	0.278	**0.587**	0.129	0.135	0.583
Qu5	0.089	0.261	**0.638**	0.028	0.082	0.550
Qu10	0.252	0.190	**0.604**	0.052	0.127	0.561
Qu11	0.247	0.204	**0.543**	0.171	0.226	0.591
Qu14	0.280	0.262	**0.528**	0.205	0.145	0.613
Qu19	0.278	0.207	**0.527**	0.069	0.192	0.547
Qu26	0.295	0.229	**0.531**	0.294	0.175	0.578
Qu36	0.290	0.271	0.214	**0.525**	0.149	0.588
Qu40	-0.050	0.237	0.221	**0.524**	0.241	0.557
Qu42	0.234	0.241	0.153	**0.607**	0.163	0.595
Qu50	0.239	0.206	-0.028	**0.469**	0.225	0.557
Qu54	0.207	0.184	0.125	0.229	**0.545**	0.537
Qu61	0.194	0.182	0.209	0.144	**0.616**	0.585
Qu65	0.049	0.149	0.136	0.140	**0.630**	0.567
Qu66	0.211	0.201	0.282	0.117	**0.498**	0.566

In summary, the new MSIS, with 29 items, had a Cronbach's alpha value of 0.95 with a standardised alpha value of 0.96. The number of participants which could be maintained in further analyses of the data was 1,085 (90.4 percent of the original sample).

Research question 1: Do Hindu-Mauritians, Muslim-Mauritians, and Creole-Mauritians differ with regard to spiritual intelligence?

A one-way Welch ANOVA was used to determine whether the spiritual intelligence (MSIS-1 score) was different for Hindu-Mauritians, Muslim-Mauritians, and Creole-Mauritians. Participants were classified into the three main groups: the Hindu-Mauritian group (n = 330), the Muslim-Mauritian group (n = 308), and the Creole-Mauritian group (n = 338). There were statistically significant differences in spiritual intelligence (MSIS-1 scores) among the different ethnic groups: Welch's F (2, 639.98) = 3.923, p = 0.02. The spiritual intelligence score (MSIS_SCORE) reached an average mean of 2 across the different ethnic groups. The Hindu-Mauritians had the highest MSIS-1 score (M = 2.36, SD = 0.70); that of

Muslim-Mauritians was slightly lower (M = 2.27, SD = 0.77), and Creole-Mauritians had the lowest score (M = 2.21, SD = 0.67). A Games-Howell post-hoc analysis showed that the mean difference between Hindu-Mauritians and Creole-Mauritians (0.15, 95 percent CI [0.023, 0.27]) was statistically significant (p = 0.015). The hypothesis that Hindu-Mauritians, Muslim-Mauritians, and Creole-Mauritians do not differ with regard to spiritual intelligence was rejected in favour of the alternative hypothesis. A one-way multivariate analysis of variance (MANOVA) of the ethnic groups on the six dimensions of MSIS-1 could not be completed due to multicollinearity. The Pearson correlations between the various dimensions varied between 0.767 and 0.888.

Research question 2: Is there a relationship between level of ethnic identity and spiritual intelligence?

The Cronbach's alpha value for the MEIM was 0.842 (the standardised alpha value was 0.850), and the reliability for the exploration subscale and the commitment subscale were 0.714 (the standardised alpha value was 0.730) and 0.834 respectively. The Pearson's coefficient was run despite the assumption violations, as the test was robust to violations. A strong positive correlation was observed between ethnic identity and spiritual intelligence, r (98) = 0.507, p < 0.0005 with ethnic identity explaining 26 percent of the variation in spiritual intelligence. Table 4. displays the correlations between the MEIM and MSIS-1 scores. The MEIM's dimensions had a moderate correlation with the MSIS-1 subscales. The hypothesis that there is no relationship between ethnic identity and spiritual intelligence was rejected in favour of the alternative hypothesis.

Measure: Variable	MSIS[3]	SA	LC	TA	S	R	QM
MEIM (n= 1054)	0.507**	0.483**	0.485**	0.489**	0.438**	0.486**	0.465**
MEIM: Exploration	0.540**	0.507**	0.499**	0.523**	0.478**	0.521**	0.503**
MEIM: Commitment	0.373**	0.362**	0.374**	0.357**	0.310**	0.353**	0.334**

Correlation was significant at the 0.01 level (2-tailed). **MSIS**: Multicultural Spiritual Intelligence Scale; **SA**: Self-awareness; **LC**: Level of Consciousness; **TA**: Transcendental awareness; **S**: Sensitivity; **R**: Resilience; **QM**: Quest for Meaning; **MEIM**: Multigroup Ethnic Identity Measure

Research question 3: Are there differences in ethnic identity in the three ethnic groups?

A one-way ANOVA was run to determine if the ethnic identity (MEIM score) was different for the diverse ethnic groups in Mauritius. The MEIM score was highest for Muslim-Mauritians (M = 2.56, SD = 0.91), followed by Hindu-Mauritians (M = 2.20, SD = 0.86); the lowest scores were those of the Creole-Mauritians (M = 2.09, SD = 0.84). The differences were statistically significant: $F_{(2, 965)} = 25.721$, $p < 0.0005$. A Tukey post-hoc analysis showed that the difference between Hindu-Mauritians and Creole-Mauritians (0.12, 95 percent CI (-0.04 to 0.27)) was not statistically significant (p = 0.198). However, the differences between Muslim-Mauritians and Hindu-Mauritians (0.36, 95 percent CI (0.20 to 0.52)) and between Muslim-Mauritians and Creole-Mauritians (0.47, 95 percent CI (0.31 to 0.64)) were statistically significant: $p < 0.0005$ for both. A one-way multivariate analysis was run to establish the effect of ethnic groups on ethnic identification. The Muslim-Mauritians had the highest score in both ethnic exploration (M = 2.64, SD = 1.20) and ethnic commitment (M = 2.49, SD = 0.91). The Hindu-Mauritians and the Creole-Mauritians had lower scores in ethnic exploration (M = 2.37, SD = 0.94; M = 2.20, SD = 0.91 respectively) and in ethnic commitment (M = 2.04, SD = 0.95; M = 1.98, SD = 0.99 respectively). The

---

[3] The MSIS and the MEIM subscales had medium correlation

differences among the ethnic groups on the combined dependent variables were statistically significant: $F (4, 1930) = 14.679$, $P < 0.0005$; Pillai's $\wedge = 0.059$; p rtial $n^2 = 0.030$. Follow-up univariate ANOVAs indicated that both ethnic exploration ($F (2, 965) = 15.918$, $p < 0. 0005$; partial $n^2 = 0.032$) and ethnic commitment ($F (2, 965) = 27.198$, $p < 0.0005$, partial $n^2 = 0.053$) were statistically significant, using a Bonferroni-adjusted α level of 0.025. A Games-Howell post-hoc analysis demonstrated that the mean difference between Hindu-Mauritians and Creole-Mauritians (0.17, 95 percent CI [-0.0084, 0.35]) was not statistically significant ($p = 0.066$). However, the mean differences between Muslim-Mauritians and Hindu-Mauritians (0.27, 95 percent CI [0.083, 0.45]) and between Muslim-Mauritians and Creole-Mauritians (0.44, 95 percent CI [0.25, 0.62]) were statistically significant ($p = 0.002$ and $p < 0.0005$ respectively). The hypothesis that the ethnic groups do not differ with regard to ethnic identification was rejected in favour of the alternate hypothesis as there was a statistically significant difference between means ($p < 0.0005$).

Research question 4: Do religious groups differ with regard to spiritual intelligence?

A one-way Welch ANOVA was carried out to establish whether the spiritual intelligence (MSIS-1 score) was different across different religious groups. The participants were from the following religious groups: Christianity ($n = 368$), Buddhism ($n = 64$), Hinduism ($n = 270$), Islam ($n = 316$), and Other ($n = 18$). MSIS-1 scores were highest for Buddhism ($M = 2.46$, $SD = 0.75$), followed by Hinduism ($M = 2.32$, $SD = 0.69$), Islam ($M = 2.28$, $SD = 0.77$), Christianity ($M = 2.21$, $SD = 0.66$), and Other ($M = 2.42$, $SD = 0.82$). However, the differences among the religious groups were not statistically significant: Welch's $F (4, 104.966) = 2.276$, $p = 0.066$. The null hypothesis is supported as there was no statistically significant difference between means ($p > 0.05$)

Research question 5: Are the differences in ethnic identity and spiritual intelligence explained by gender?

A one-way multivariate analysis of variance was run to ascertain the difference in ethnic identity and spiritual intelligence according to gender. The two measures used were the MEIM score and the MSIS-1 score. Of the participants, 492 were male and 538 were female. Male participants scored higher on the MEIM (M = 2.29, SD = 0.88) than female participants (M = 2.25, SD = 0.86). However, female participants scored higher on the MSIS-1 (M = 2.29, SD = 0.70) than their male counterparts (M = 2.26, SD = 0.72). The differences in gender on the combined dependent variables were not statistically significant: $F (2, 1026) = 0.952$, $p = 0.386$; Wilks' $\wedge = 0.998$; p r i l $n^2 = 0.002$. As no statistically significant difference was found between means ($p > 0.05$), the null hypothesis is supported.

Study 2

Participants

Following the EFA in Study One, the revised version of MSIS-2 was tested in a second study. In this study, 303 survey respondents (N = 303) (49.2 percent males and 50.8 percent females) completed the 29-item questionnaire (see appendix B). In the second study, companies were randomly selected in odd numbers. Instead of simply mailing the questionnaire to the selected organisations, the researcher scheduled an appointment with the head of each organisation to explain the purpose of the study and to ensure that the seriousness of the study was well understood at the beginning. Upon the approval of the manager or the director, copies were left for completion. The questionnaires were collected within a month.

Measures: MSIS-2

All participants completed the MSIS-2 survey (see appendix B) using either the Web-based format or the paper version. Apart from the 29-item Likert scale section, seven items on gender, age, marital status, educational and occupational background, religion, and ethnic cultural origin were mentioned in the questionnaire's demographic part. The instructions about the survey were provided at the beginning. The aims of this second study were to develop a psychometric evaluation of the MSIS-2 and to test its construct validity on a random sample of Mauritians citizens (Harrington, 2009; Hoyle, 2000). IBM SPSS version 22 and IBM SPSS Amos version 22 were the software used for data entry and research analysis.

Procedures: Survey 2

Questionnaires which were emailed to respondents represented 40.6 percent of the sample. Participants who responded to the web-based version represented only 17.5 percent of the sample. The web-based format was available online for a three-month period. The participants who responded using the paper version represented 41.9 percent of the sample. Of the 303 participants, 34.7 percent denoted Christianity as their religion; 40.3 percent Hinduism, 23.4 percent Islam while Buddhism and Others represented 1 percent and less than 1 percent respectively. The "others" category denoted all participants that did not identify with any of the given categories. Although the initial sample was 450 participants, only 67 percent responded to either the emailed Web-based questionnaire or the paper version.

Results of study 2: Confirmatory factor analysis of the spiritual intelligence scale

Descriptive statistics and response distributions were assessed for the 29 items (as shown in Appendix A). The mean response of all items was higher than 2.2, which was equivalent to the response of 'moderately true of me'. The z-score for skewness and kurtosis was ±2.58, at a 0.01 significance level. All items were significant in terms of skewness and kurtosis except item 5, which had slight negative skewness, and items 12, 13, 21 and 22, which were negatively kurtosed. However, all items were retained for further analysis. The Cronbach's alpha value for the 29-item set was 0.687 (standardised alpha = 0.686), indicating a good but lower internal consistency and reliability for the whole item pool. The inter-item correlation matrix indicated a correlation coefficient within the range of 0.15 and 0.50 (Clark & Watson, 1995) with the exception of items 1, 15, 17, and 18.

The Analysis of Moment Structures (AMOS) graphics program version 22 was used for the CFA (IBM SPSS, 2013). In the hypothesised model, the five factors of the MSIS-2 (self-mastery, transcendental awareness, spiritual sensitivity, resilience, and existential quest) were the latent variables, and the 29 items were the measured variables. Items 1 to 7 served as indicators of the self-mastery subscale; items 8 to 11 were indicators of the transcendental awareness subscale; items 12, 17, 18, and 19 functioned as indicators of spiritual sensitivity; items 13, 14, 15, and 23 operated as indicators of the resilience subscale; and items 16, 20, 21, 22, 24, 25, 26, 27, 28, and 29 served as indicators of the existential quest subscale (see Appendix B). The five factors were assumed to co-vary with each other. The dataset covered responses from 303 participants (149 males and 154 females), and there were no missing data in the set. The new MSIS-2 scores were normally distributed for both males, with a skewness of -0.101 (SE = 0.199) and kurtosis of -0.881 (SE = 0.395), and females, with a skewness of -0.098 (SE = 0.195) and kurtosis of -0.684 (SE = 0.389). There was no multicollinearity, as examined by the Pearson correlation (see Table 5.). There was a medium correlation among the variables. The Maximum correlation was r = 0.400, p < 0.0005. There were 29 measured variables, and therefore 29(29+1)/2 = 435 data points or sample moments and 102 parameters to be estimated in the hypothesised model. Thus, the model was identified.

Table 5. Correlations among latent variables for study 2 (N = 286)

Correlations among the dimensions

	Self-Awareness	Transcendental Awareness	Spiritual Sensitivity	Resilience	Existential Quest
Self-Awareness	1	$.286^{**}$	$.201^{**}$	$.279^{**}$	$.348^{**}$
Transcendental Awareness	$.286^{**}$	1	$.283^{**}$	$.212^{**}$	$.390^{**}$
Spiritual Sensitivity	$.201^{**}$	$.283^{**}$	1	.110	$.373^{**}$
Resilience	$.279^{**}$	$.212^{**}$	.110	1	$.400^{**}$
Existential Quest	$.348^{**}$	$.390^{**}$	$.373^{**}$	$.400^{**}$	1

**. Correlation was significant at the 0.01 level (2-tailed).

Measurement and structural invariances of the MSIS-2

In a multicultural study, it is important to fulfill three levels of measurement invariances (Comşa, 2010). The chi square test for differences suggested that the factor structure was not invariant among the three ethnic groups, $\chi 2$ (1101, N = 279) = 1248.29, p = 0.001. This indicated that the interpretation of the dimensions of SQ was not the same among the ethnic groups. Other fit indexes supplemented the chi square test (Hu & Bentler, 1999). The RMSEA reported a value of 0.022, indicating a good fit. The CMIN/DF had a value of 1.134, supporting the model. The Pclose value was 1.000, suggesting a good fit (Comşa, 2010). The metric invariance requires configural invariance and that the factor loadings are invariant for the three ethnic groups (Comşa, 2010). The author noted that metric and scalar invariances were needed for a minimum of two items per dimensions for meaningful mean comparison. Based on this principle, only self-mastery, transcendental awareness and existential quest met the criteria (Table 6.). The fit indices indicated that the metric invariance model should not be rejected (RMSEA = 0.024, Pclose = 1.000 and the CMIN/DF = 1.162). Scalar invariance requires metric invariance and that the three samples have invariant intercept scores. A model comparison revealed nonsignificant scalar invariance, $\chi 2$ (58, N = 279) = 63.978, p > 0. 005. The model fit also suggested that the scalar invariance model should not be rejected (RMSEA= 0.024, Pclose = 1.000 and the CMIN/DF = 1.159) as shown in table 7.

Table 6. Comparing the dimensions of MSIS-2 (N = 279) for the three ethnic groups for metric invariance

Dimensions	CMIN	df	P value	Requirement p > 0.05 for metric invariance
Self-mastery	24.193	14	0.043	Close to invariance
Transcendental awareness	10.434	8	0.236	Yes
Spiritual sensitivity	20.548	8	0.008	No
Resilience	27.071	8	0.001	No
Existential quest	30.569	20	0.061	Yes

Table 7. *Goodness-of-fit statistics for measurement invariance among the three ethnic groups (N = 279)*

Invariance type	CMIN/DF	CFI	RMSEA	PCLOSE	AIC
Criteria	< 2	> 0.95	< 0.08	> 0.5	-
Configural	1.134	0.673	0.022	1.000	1831.715
Metric	1.162	0.584	0.024	1.000	1820.511
Scalar	1.159	0.574	0.024	1.000	1768.488

The test for residual invariance is the final step in examining measurement invariance. There was a nonsignificant difference in item residual variances among groups, $\chi 2$ (58, N = 279) = 40.338, p > 0. 005, see table 8. The three ethnic groups had the same item residual variances. Structural invariance relates to how the dimensions of the MSIS-2 are distributed and related in the separate populations such as the three ethnic groups (Templin, 2012). The test for the structural invariance was to determine if the causal relationships functioned the same way across the three ethnic groups. The nonsignificant difference was observed across the three ethnic groups, $\chi 2$ (10, N = 279) = 11.319, p > 0. 005.

Table 8. Model comparisons for measurement and structural invariances across the three ethnic groups (N = 279)

	CMIN	DF	P value	Requirement for invariance (p > 0.05)
Metric	80.796	46	0.001	No
Scalar	63.978	58	0.275	Yes
Residual	40.338	58	0.962	Yes
Structural	11.319	10	0.333	Yes

Modal fit measures for the factor model of the MSIS-2

Table 9. shows the parameter estimate or path coefficients for each measured variable for the five-factor model. Only 14 measured variables reached significance (p < 0.001 or p < 0.05). Items 2, 5, 8, 12, and 13 were significant predictors of their respective factors, as they reached practical significance—that is, they exceeded the 0.3 criterion (Meyers et al., 2006). Items 3 and 7 were both 0.30, which could be accepted as having practical significance. Although many fit measures reported an acceptable model, this could be because some measured paths had very high coefficients (Meyers et al., 2006).

Table 9. Parameter estimate for five-factor model (N = 286)

Measured variables	Latent variables	Parameter estimate	Measured variables	Latent variables	Parameter estimate
Qu1	SM	.033	Qu13	R	.557
Qu2	SM	.398	Qu14	R	.296**
Qu3	SM	.268	Qu15	R	.123
Qu4	SM	.225	Qu23	R	.355**
Qu5	SM	.439	Qu16	EQT	.188
Qu6	SM	.237	Qu20	EQT	.357*
Qu7	SM	.278	Qu21	EQT	.350*
Qu8	TA	.378	Qu22	EQT	.133
Qu9	TA	.386**	Qu24	EQT	.218*
Qu10	TA	.402**	Qu25	EQT	.085
Qu11	TA	.422**	Qu26	EQT	.175*
Qu12	SS	0.490	Qu27	EQT	.303*
Qu17	SS	0.071	Qu28	EQT	.206*
Qu18	SS	0.213*	Qu29	EQT	.224*
Qu19	SS	0.529**			

SM = self-mastery; TA= Transcendental awareness; SS = spiritual sensitivity; R = resilience; EQT = Existential quest. * p < 0.05, ** p < 0.001

The chi square had a value of 430.222 (367, N = 286), p = 0.013, which was significant and indicated an improper fit between the proposed model and the observed data. Because the chi square was limited and sensitive to sample size, an alternative had to be explored, namely the relative chi-square or χ^2/df ratio (King, 2008; Moss, 2009). In this study, the normed chi-square was 430.222/367 = 1.17, which was below the maximum value of 2.0 for good fit. The CFI and the NFI were measures of relative fit, contrasting the hypothesised model with the null model by using suitable values of 0.95 (Hu & Bentler, 1999, as cited in Meyers et al., 2006). Both the CFI and NFI produced values of 0.549 and 0.212 respectively, which were both below 0.95 and thus suggested a poor fit for the model. The GFI was 0.896 below the cut-off value, while the AGFI was 0.877 above the minimum value. The SRMR was 0.0586, which indicated a good fit. The RMSEA was 0.025, showing good fit.

The comparative fit indices for five-factor, four-factor, and three-factor models are shown in Table 10. In the four-factor model, the first three factors were self-mastery, transcendental awareness, and the existential quest, while the spiritual sensitivity subscale encompassed the resilience scale. In the three-factor model, self-mastery and transcendental awareness were combined into one factor named self-mastery, and spiritual sensitivity and resilience were similarly united in one factor called spiritual sensitivity; the third factor remained the existential quest. There was no substantial difference in terms of fit measures for the three models. Therefore, the five-factor model was chosen as the best fit model of the data.

Table 10. Comparative fit indices for the five-factor, four-factor, and three-factor models (N = 286)

	χ^2	χ^2/df ratio	CFI	NFI	GFI	AGFI	SRMR	RMSEA	*df*
Criteria set	- p> 0.05	< 2	> 0.95	> 0.95	> 0.95	> 0.85	< 0.08	< 0.08	-
Value for **(a) 5-factor** **model**	430.222, p= 0.013	1.172	0.549	0.212	0.896	0.877	0.0586	0.025	367
(b) 4-factor **model**	445.365, p= 0.005	1.200	0.469	0.184	0.892	0.874	0.0598	0.027	371
(c) 3-factor **model**	445.518, p= 0.006	1.19	0.490	0.184	0.892	0.875	0.0598	0.026	374

Discussion

The current literature describes spiritual intelligence (SQ) in populations which do not reflect the Mauritian reality. This dearth of knowledge on SQ in a population marked by ethnic lines motivated this research. The main ethnic groups which were the focus of this study were Hindu-Mauritians, Muslim-Mauritians, and Creole-Mauritians. The primary purpose of the present study was to examine the nature of SQ and its link to ethnic identification in multi-ethnic Mauritius.

The MSIS compared to other measures

Although an average correlation was found between spiritual intelligence and a religious orientation scale (NIRO: extrinsic), the subscales had a low correlation with SQ, which confirmed MSIS's construct validity. Strong positive relationships between spiritual intelligence and private self-consciousness (PSC) also highlight this convergent validity. Findings by Kumari and Sharma (2011) confirmed the relationship between the two variables, and indicated that private self-consciousness is a high predictor of SQ. The low correlation between the MSIS and the social desirability scale (BIDR) and its subscales confirmed their discriminant validity. This low correlation also means that the MSIS measured a different construct. Social desirability is respondents' propensity to favour items in response to normative demands (Ellingson, Smith & Sackett, 2001, as cited in Li & Bagger, 2007). The BIDR measured the conscious and unconscious exaggerations that distort responses. A low correlation was also observed between BIDR and all other measures. The comparison between the MSIS and the meaning in life questionnaire (MLQ) (Steger et al. 2006) showed a very good relationship, validating the MSIS's construct. The correlation between the presence of meaning and SQ was stronger than the correlation between the search for meaning and SQ. Similar findings were observed by King and DeCicco (2009), who claimed that presence of meaning was a consequence of the ability to create meaning, and thus concurrent validity was supported. The meaning dimension as related to SQ implied that the quest phase preceded the achievement of meaning in life. Therefore, the presence of meaning indicated a more advanced ability to generate meaning. The items included in the MSIS were meant to measure this advanced ability. The strong link between spiritual intelligence and resilience supported convergent validity. Shahbakhsh and Moallemi (2013) confirmed a positive link between SQ and resilience using King's SQ inventory and CD-RISC to assess resilience. Their work also confirmed Zohar and Marshall's (2005) initial proposition to include resilience as a potential component of SQ. King and DeCicco's (2009) study also revealed an equivalent transcendental awareness construct that validated the transcendental awareness component in the MSIS.

Ethnic differences in spiritual intelligence

Among the three ethnic groups concerned, the Hindu-Mauritians attained the highest score in SQ. The only significant difference in spiritual intelligence observed was between the Hindu-Mauritians and the Creole-Mauritians. The Hindu Mauritians include Hindus, Tamils, Telegus, and Marathis (Eriksen, 1999) who share the common thread of Hindu philosophy but differ in terms of language, temple structure, presiding divinity, and culinary culture. The Eastern perspective emphasises transcendence to rediscover a higher level of awareness and consciousness, which were major themes of the MSIS-1 as initially constructed (Wilber, 1993). Furthermore, the goal of self-realisation, which corresponded to the liberation of the soul in Hinduism, involved the complete transformation of the person for the experience of the real self (Viljoen, 2003). This transformation in turn concerned a quest for meaning for a higher purpose that intersected with the development of a spiritually intelligent individual. Knowledge of the self and the absolute are prerequisites for moksha, the ultimate deliverance, which is also gained through a search for meaning (Rambachan, 2000).

Ethnic identification showed a significant yet moderate relationship with spiritual intelligence. This validates the MSIS, as both ethnic identity and spiritual intelligence development necessitate a search and an inner quest that culminates in an understanding or sense of self. The influence of ethnic identification, actively developed from a time of search and questioning, is substantial in safeguarding oneself from mere adherence to a group and manipulation by others, in ensuring a confident constructive sense of one's identity as part of a group, and in fostering tolerance of other groups (Phinney, 1996). Unlike an ethnic identity, SQ involves a continuous quest for meaning and development in the several areas of SQ. It is a herculean task in the real world which climaxes in a state which is comparable to Maslow's (1943) self-realisation theory. The exploration dimension was found to have a stronger significant link with SQ than the commitment component: the element of exploration or quest was present in both building an ethnic identity and the spiritual quest. The moratorium, a period of uncertainty, self-questioning, and 'ambivalence' (Phinney et al., 2007, p. 479) forges character and builds the SQ of the individual.

Gender and spiritual intelligence

Regarding the positive correlation between the level of ethnic identity and spiritual intelligence, gender was not a moderating variable in this relationship; rather, it was found that SQ and ethnic identity were positively linked irrespective of gender. The more spiritually intelligent a person is, the higher their level of ethnic identification and vice versa. Male participants scored slightly higher on the MEIM than females, confirming the results of previous investigations of gender differences in ethnic identity (Phinney, 1992). On the MSIS scale, females slightly outscored males, which was also supported by the results of several other studies (Allah Du, Mazdarani & Ghasemian, 2013; Azizi & Zamaniyan, 2013). While Allah Du et al. (2013) did not provide any details regarding the assessment tool they used in their research, Azizi and Zamaniyan (2013) used King's (2008) spiritual intelligence inventory in their study. Conversely, Nazam's (2014) small-sample study of adolescents in India found males outperformed females in SQ. Nazam (2014) also used King's (2008) scale, which considered critical existential thinking, personal meaning production, transcendental awareness, and conscious state expansion as components of spiritual intelligence. Though the two authors used the same measure, the different results could mean a difference in the sample or participants in the study. However, the fact that Nazam's (2014) study used a small sample (only 60 participants) could also explain the divergent results. Azizi and Zamaniyan (2013) and Allah Du et al. (2013) each used 120 participants, which was also insufficient to generalise the results. Differences in how the components of SQ were measured could also explain the gender differences. King's (2008) measure had similar components to the MSIS. However, the MSIS had more dimensions, such as sensitivity, self-awareness, and resilience, which could explain the amplified difference.

Ethnic identifications of the major ethnic groups

The difference in ethnic identification was significant between Muslim-Mauritians and Hindu-Mauritians and between Muslim-Mauritians and Creole-Mauritians. In short, the substantial difference between Muslim identity and the others validates the MEIM measure. Muslim-Mauritians had the highest score in both the ethnic exploration and ethnic commitment subscales, which meant that they obtained the highest score with regard to both the affective and cognitive aspects. The Indo-Mauritian Muslims, a minority group who represent 17 percent of the

Mauritian population (Eisenlohr, 2006; Eriksen, 1999), are perceived as successful traders, due in part to their prosperous ancestral Gujarati origins. On one hand there are the noticeably wealthy and influential Memon and Surtee Muslims, while on the other hand, there are the other Muslim groups from modest, unskilled backgrounds, the descendants of indentured servants (Eisenlohr, 2006). Social identity theory hints that group membership becomes salient in circumstances where people are part of a minority group and perceive some degree of threat (Syed & Juang, 2014). The conflicts within the Muslim group, which have given rise to subgroups, could have reinforced the subgroup identification, which could also explain the stronger degree of identification reported in the present study. Juang and Syed (2010) related similar findings—that is, ethnic minorities described a stronger sense of ethnic identity than the White participants. The Creole-Mauritians obtained the lowest score. Ng Tseung-Wong and Verkuyten's (2010) study also described Creoles' relatively negative evaluation of their own group. Phinney (1991, as cited in Mossakowski, 2003) claimed that ethnic identification was a 'sense of pride' (p. 318) and included participation in ethnic rituals and cultural adherence to the ethnic group, but this was not found to be the case for the Creole-Mauritian group. The constructivist perspective paints ethnicity as flexible and conditional upon many variants (Isaacs-Martin, 2014; see also Anderson, 2001), so it could be assumed that Creole-Mauritians refer to their identity as a 'creolization' (Erasmus, 2001 as cited in Isaacs-Martin, 2014, p. 57) moulded by slavery and discrimination in the absence of a history prior to colonialism (see also Eriksen, 1999). The Creole-Mauritians' low score on ethnic identity questions the established ethnic groups in Mauritius. Furthermore, the constructivist viewpoint is inadequate in this instance and should be taken with other perspectives. Although the socio-historical commonalities exist, the Creole-Mauritians do not identify themselves as an ethnic group but harbour a subjective view of their ethnic statuses. From an instrumentalist outlook, being the least privileged and marginalized group can also be a reason behind the lack of enthusiasm to identify with a disadvantaged group. A circumstantialist approach would look at the 1999 riots, that reawakened the ethnic Creole bonds, as a reaction to the felt oppressive treatment. This perspective explains how structured inequality can contribute to the temporary phase of ethnic group formation. Boswell (2006) described the Creole-Mauritians as a marginalised group who were deprived of an ethnic identity and lacked an 'ancestral culture' (p. 693), being conscious instead of their descent from slaves and the legacy thereof (Ng Tseung-Wong & Verkuyten, 2015). The Creole community is afflicted by low educational attainment, acute unemployment, constant poverty, a low standard of living, social issues, 'political marginalisation', and an absence of solidarity among themselves which is called *le malaise creole* (Boswell, 2006, p. 2; see also Jeffery, 2007, 2010). Proponents of pluralism propose that recognition of and consideration for valued subgroup identities will have a unifying impact on the social system, and would make people of different backgrounds feel accepted (Huo & Molina, 2006; see also Molina, Phillips & Sidanius, 2015).

Conclusion

This research focused on the study of spiritual intelligence among three main ethnic groups in Mauritius. A Multicultural Spiritual Intelligence Scale (MSIS) was developed, taking into consideration the African and Eastern perspective in a broad item distribution for a universal and comprehensive spiritual intelligence scale. The MSIS was used to explore ethnic identification and ethnic differences in Mauritius. The study adds to the literature on spiritual intelligence in the Mauritian diversity context by considering ethnic, gender, and religious differences. Two studies were conducted for this purpose. In summary, ethnic groups differed with regard to SQ; a most significant difference was between Hindu-Mauritians and Creole-Mauritians. Ethnic identity was significantly correlated with SQ. The various ethnic groups were also found to differ in their ethnic identification in terms of exploration and commitment. Compared to the commitment dimension of the MEIM, the exploration dimension was more strongly correlated with the subscales of the MSIS. However, there were no statistical significant differences among the religious groups in terms of SQ. Concerning the positive relationship between the level of ethnic identity and spiritual intelligence, gender did not play a moderating role. Study Two confirmed the factor structure observed in Study One, and a five-factor model was chosen as the model which best fit the data.

Limitations and Recommendations for Further Research

The study did not consider the minority groups or dual identities like the Sino-Mauritians and the Franco-Mauritians. Furthermore, the research overlooked participants who had no ethnic identifications. The Creole-Mauritians' low score on ethnic identification questioned the group as an ethnic one, from a constructivist perspective. This research could be a ground work for a more in-depth qualitative survey on the ascribed Creole-Mauritian. The present research study centred on three main ethnic groups: Hindus, Muslims, and Creoles. Further studies on ethnic minorities such as Sino-Mauritians and Franco-Mauritians with regard to spiritual intelligence would not only add to the knowledge on the subject but also help in the better understanding of ethnic differences in SQ. Moreover, for generalisation purposes, the MSIS could be used on different populations (such as a South African population, for example) to confirm its psychometric properties. Based on the findings of the present study, it would be in the nation's interest to work on a more inclusive multicultural policy for the historically marginalised Creole-Mauritians (Huo & Molina, 2006). The differences in SQ among the ethnic groups can be easily misinterpreted and nurture stereotypes and social discrimination, which are not intended in this research.

Declaration of Conflicting Interests

The authors declared no potential conflicts of interests with respect to the authorship and/or publication of this article.

Funding

This study was supported by a doctoral bursary from the University of South Africa.

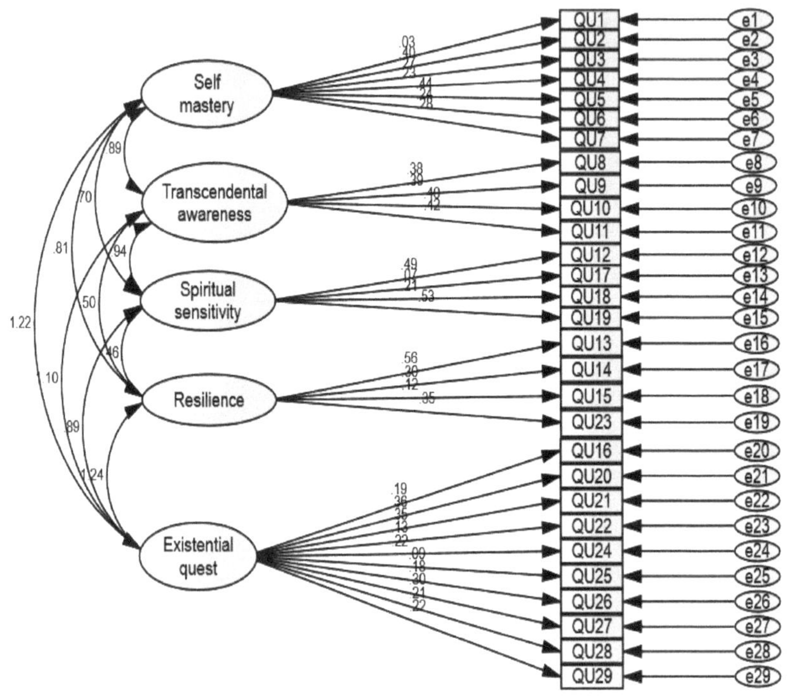

Figure 1 **Path diagram for the hypothesised five-factor model.**

References

Addison, J., & Hazareesingh, K. (1999). *A new history of Mauritius*. Rose Hill : Editions de L'Ocean Indien.

Allah Du, S., Mazdarani, S., & Ghasemian, D. (2013). The relationship between spiritual intelligence and stress among teachers. *Journal of Social Issues & Humanities*, 1(6), 58-60.

American Psychological Association. (2017). *The road to resilience*. Retrieved July, 2014, from http://www.apa.org/helpcenter/road-resilience.aspx

Amram, Y., & Dryer, D. C. (2008). The integrated spiritual intelligence scale (ISIS): development and preliminary validation. *116th Annual Conference of American Psychological Association*. Retrieved from: http://www.yosiamram.net/docs/ISIS_APA_Paper_Presentation_2008_08_17.pdf

Anderson, A. B. (2001). The complexity of ethnic identities: a postmodern reevaluation. *Identity: an International Journal of Theory and Research*, 1(3), 209-223. doi: 10.1207/S1532706XID0103_02

Appel, J., & Appel, D. K. (2009). Mindfulness: implications for substance abuse and addiction. *International Journal of Mental Health Addiction*, 7, 506-512. doi: 10.1007/s11469-009-9199-z

Armstrong. T (2009). *Multiple intelligences in the classroom*. Virginia: ASCD

Atkinson, P. A., Martin, C. R., & Rankin, J. (2009). Resilience revisited. *Journal of Psychiatric and Mental Health Nursing,* 16, 137-145.

Azizi, M., & Zamaniyan, M. (2013). The relationship between spiritual intelligence and vocabulary learning strategies in EFL learners. *Theory and Practice in Language Studies*, 3(5), 852-858. doi: 10.4304/tpls.3.5.852-858

Batson, C. D., & Schoenrade, P. A. (1991). Measuring Religion as Quest: 1) Validity Concerns. *Journal for the Scientific Study of Religion, 30*(4), 416. doi:10.2307/1387277

Boswell, R. (2006). *Le malaise Créole : ethnic identity in Mauritius*. New York : Berghahn Books

Burrun, B. (2002). *Histoire des religions des Îles Maurice et Rodrigues*. Mauritius : Éditions Le Printemps Ltée.

Calvillo, J. E., & Bailey, S. R. (2015). Latino religious affiliation and ethnic identity. *Journal for the Scientific Study of Religion*, 54(1), 57-78.

Carpooran, A. (2003). *Île Maurice : des langues et des lois*. Paris : L'Harmattan

Clark, L. A., & Watson, D. (1995). Constructing validity: Basic issues in objective scale development. *Psychological Assessment, 7*(3), 309-319. doi:10.1037//1040-3590.7.3.309

Clinton, J. (2008). Resilience and recovery. *International Journal of Children's Spirituality, 13*(3), 213-222. doi:10.1080/13644360802236474

Comşa, M. (2010). How to compare means of latent variables across countries and waves: Testing for invariance measurement. An application using Eastern European societies. *Sociológia*, 42(6), 639-669.

Connor, K. M., & Davidson, J. R. T. (2011). *Connor-Davidson resilience scale*. Retrieved from http://www.cd-risc.com

Creswell, J. W. (2012). *Educational research: Planning, conducting, and evaluating quantitative and qualitative research*. Boston: Pearson Education Inc.

Davoudi, S. (2013). On resilience. *DisP-The Planning Review*, 49(1), 4-5. doi:10.1080/02513625.2013.799852

DeHoff, S. L. (1998). In search of a paradigm for psychological and spiritual growth: implications for psychotherapy and spiritual direction. *Pastoral Psychology*, 46(5), 333-346

Dillen, A. (2012). The resiliency of children and spirituality: a practical theological reflection. *International Journal of Children's Spirituality*, 17(1), 61-75. doi: 10.1080/1364436X.2012.670616

Eisenlohr, P. (2006). The politics of diaspora and the morality of secularism: Muslim identities and Islamic authority in Mauritius. *Journal of the Royal Anthropological Institute*, 12, 395-412.

Emmons, R. A. (2000), Spirituality and intelligence: Problems and prospects. *International Journal for the Psychology of Religion, 10*(1), 57-64. Retrieved from http://dx.doi.org/10.1207/S15327582IJPR1001_6

Eriksen, T. H. (1999). *Tu dimunn pu vini kreol: The Mauritian creole and the concept of creolization*. Paper presented at the Creolization Seminar, Oxford.

eSurvey Creator. (2015). Creation and evaluation of surveys [Online survey]. Retrieved from https://www.esurveycreator.com

Feist, J., & Feist, G. J. (2002). *Theories of personality*. New York: Mc Graw-Hill.

Ferrari, M. (1998). Being and becoming self-aware. In M. Ferrari, & R. J. Sternberg (Eds.), *Self-awareness: Its nature and development* (pp. 387-422). New York: The Guilford Press.

Francis, L. J. (2007). Introducing the new indices of religious orientation (NIRO): Conceptualization and measurement. *Mental Health, Religion & Culture*, 10(6), 585-602.

García Coll, C., Crnic, K., Lamberty, G., Wasik, B. H., Jenkins, R., Vázquez García, H., & McAdoo H. P. (1996). An integrative model for the study of developmental competencies in Minority Children. *Child Development*, 67, 1891-1914.

Gardner, H. (1999). *Intelligence reframed: Multiple intelligences for the 21ˢᵗ Century*. New York: Basic Books.

Gardner, H. (2006). *Multiple intelligences: New horizons*. New York: Basic Books.

Gardner, H. (2011). *Frames of mind: The theory of multiple intelligences*. New York: Basic Books.

Government Portal (2015). *Explore Mauritius: Geography & people*. Retrieved from http://www.govmu.org/

Greenway, A. P., Phelan, M., Turnbull, S., & Milne L. C. (2007). Religious coping strategies and spiritual transcendence. *Mental Health, Religion & Culture*, 10(4), 325-333. doi: 10.1080/13694670600719839

Greig, R. (2003). Ethnic identity development: Implications for mental health in African-American and Hispanic adolescents. *Issues in Mental Health Nursing*, 24, 317-331. doi: 10.1080/01612840390160810

Harrington, D. (2009). *Confirmatory factor analysis*. New York: Oxford University Press.

Hay, D., & Nye, R. (2006). *The spirit of the child*. London: Fount.

Helms, J. E. (2007). Some better practices for measuring racial and ethnic identity constructs. *Journal of Counseling Psychology*, 54(3), 235-246. doi: 10.1037/0022-0167.54.3.235

Holas, P., & Jankowski, T. (2013). A cognitive perspective on mindfulness. *International Journal of Psychology*, 48(3), 232-243. doi:10.1080/00207594.2012.658056

Hoyle, R. H. (2000). Confirmatory factor analysis. In H. E. A. Tinsley, & S. D. Brown (Eds.), *Handbook of applied multivariate statistics and mathematical modeling* (pp. 465-492). San Diego: Academic Press.

Hu, L., & Bentler, P. M. (1999). Cutoff criteria for fit indexes in covariance structure analysis: Conventional criteria versus new alternatives. *Structural Equation Modeling: A Multidisciplinary Journal,* 6(1), 1-55.

Huo, Y. J., & Molina, L. E. (2006). Is pluralism a viable model of diversity? The benefits and limits of subgroup respect. *Group Processes & Intergroup Relations*, 9(3), 359-376. doi: 10.1177/1368430206064639

IBM (2013). *IBM SPSS statistics 22 brief guide*. New York: IBM Corporation.

Isaacs-Martin, W. (2014). National and ethnic identities: Dual and extreme identities amongst the coloured population of Port Elizabeth, South Africa. *Studies in Ethnicity and Nationalism*, 14(1), 55-73.

Jeffery, L. (2007, November). Rosabelle Boswell, Le malaise Créole: ethnic identity in Mauritius. [Review of the book *Le malaise creole: ethnic identity in Mauritius,* by R. Boswell]. *Africa*, 77(4), 611-612. doi: 10.3366/afr.2007.77.4.611

Jeffery, L. (2010). Creole festivals and Afro-Creole cosmopolitanisms in Mauritius. *Social Anthropology*, 18(4), 425-432. doi: 10.1111/j.1469-8676.2010.00126.x

Juang, L., & Syed, M. (2008). Ethnic identity and spirituality. In R. M. Lerner, R. W. Roeser, W. Robert, & E. Phelps (Eds.), *Positive youth development and spirituality: From theory to research (pp.262-284)*. West Conshohocken, PA, US: Templeton Foundation Press.

Jungers, C. M., Gregoire, J., & Slagel L. (2009). Racial/ethnic identity among Creole people in Mauritius. *Journal of Psychology in Africa*, 19(3), 301-308. doi: 10.1080/14330237.2009.10820295

Jungers, C. M., & Gregoire, J. (2010). Mauritian Creole identity development and influences of the Catholic Church. *Journal of Ethnographic & Qualitative Research*, 5, 84-98.

Kasen, S., Wickramaratne, P., Gameroff, M. J., & Weissman, M. M. (2012). Religiosity and resilience in persons at high risk for major depression. *Psychological Medicine, 42,* 509-519. doi:10.1017/S0033291711001516

King, D. B. (2008). *Rethinking claims of spiritual intelligence: A definition, model, and Measure* (Unpublished master's thesis). Trent University, Peterborough, ON.

King, D. B., & DeCicco, T. L. (2009). A viable model and self-report measure of spiritual intelligence. *International Journal of Transpersonal Studies, 28*, 68-85. Retrieved from http://www.transpersonalstudies.org

King, P. E., Clardy, C. E., & Ramos, J. S. (2014). Adolescent spiritual exemplars: Exploring spirituality in the lives of diverse youth. *Journal of Adolescent Research, 29*(2), 186-212. doi: 10.1177/0743558413502534

Kumari, P., & Sharma, G. (2011, October). *Consciousness as a predictor of spiritual intelligence.* Paper presented at the Indian Psychological Science Congress, Chandigarh, India.

Li, A., & Bagger, J. (2007). The balanced inventory of desirable responding (BIDR): A reliability generalization study. *Educational and Psychological Measurement*, 67(3), 525-544. doi: 10.1177/0013164406292087.

Lund Research Ltd (2013). *Laerd statistics.* Retrieved from https://statistics.laerd.com

Martinez, R. O., & Dukes, R. L. (1997). The effects of ethnic identity, ethnicity, and gender on adolescent well-being. *Journal of Youth and Adolescence*, 26(5), 503- 516.

McKinlay, A., & McVittie, C. (2011). *Identities in context: Individuals and discourse in action.* West Sussex: Wiley-Blackwell.

Meyers, L. S., Gamst, G., & Guarino, A. J. (2006). *Applied multivariate research: Design and interpretation.* Thousand Oaks: Sage Publications.

Molina, L. E., Phillips, N. L., & Sidanius, J. (2015). National and ethnic identity in the face of discrimination: Ethnic minority and majority perspectives. *Cultural Diversity and Ethnic Minority Psychology*, 21(2), 225-236. doi: 10.1037/a0037880

Moss, S. (2009, April 27). Fit indices for structural equation modeling [Web log post]. Retrieved from http://www.psych-it.com.au/Psychlopedia/article.asp?i

Mossakowski, K. N. (2003). Coping with perceived discrimination: Does ethnic identity protect mental health? [Special Issue]. *Journal of Health and Social Behavior*, 44(3), 318-331. Retrieved from: http://www.jstor.org/stable/1519782

Nasel, D. D. (2004). *Spiritual orientation in relation to spiritual intelligence: A new consideration of traditional Christianity and new age/individualistic spirituality.* (Unpublished doctoral dissertation). University of South Australia, Australia.

Nazam, F. (2014). Gender difference on spiritual intelligence among adolescents. *Indian Journal of Applied Research*, 4(11), 423-425.

Ng Tseung-Wong, C., & Verkuyten, M. (2010). Intergroup evaluations, group indispensability and prototypicality judgments: A study in Mauritius. *Group Processes & Intergroup Relations, 13*(5), 621-638. doi: 10.1177/1368430210369345

Ng Tseung-Wong, C., & Verkuyten, M. (2015). Multiculturalism, Mauritian style: Cultural diversity, belonging, and a secular state. *American Behavioral Scientist*, 59(6), 679-701. doi: 10.1177/0002764214566498

Pahl, K., & Way, N. (2006). Longitudinal trajectories of ethnic identity among urban Black and Latino Adolescents. *Child Development*, 77(5), 1403-1415.

Paulhus, D. L. (1991). Measurement and control of response bias. In J. P. Robinson, P. R. Shaver, & L.S. Wrightsman (Eds.), *Measures of personality and social psychological attitudes* (pp.17-59). San Diego, CA: Academic Press Retrieved from: http://www.sjdm.org/dmidi/Balanced_Inventory_of_Desirable_ Responding.html

Phinney, J. S. (1990). Ethnic identity on adolescents and adults: Review of research. *Psychological Bulletin*, 108 (3), 499-514.

Phinney, J. S. (1992). The multigroup ethnic identity measure: A new scale for use with diverse groups. *Journal of Adolescent Research*, 7(2), 156-176. doi: 10.1177/074355489272003
Phinney, J. S. (1996). Understanding ethnic diversity: The role of ethnic identity. *American Behavioral Scientist*, 40(2), 143-152. doi: 10.1177/0002764296040002005

Phinney, J. S., & Ong, A. D. (2007). Conceptualization and measurement of ethnic identity: Current status and future directions. *Journal of Counseling Psychology*, 54(3), 271-281. doi: 10.1037/0022-0167.54.3.271

Phinney, J. S., Jacoby, B., & Silva, C. (2007). Positive intergroup attitudes: The role of ethnic identity. *International Journal of Behavioral Development*, 31(5), 478-490. doi: 10.1177/0165025407081466

Ramadan, T. (2012). *The quest for meaning*. London: Penguin.

Rambachan, A. (2000). Hinduism and the encounter with other faiths. *Global Dialogue*, 65-73.

Richardson, G. E. (2002). The metatheory of resilience and resiliency. *Journal of Clinical Psychology*, 58(3), 307-321. doi: 10.1002/jclp.10020

Scheier, M. F., & Carver, C. S. (1985). The self-consciousness scale: A revised version for use with general populations. *Journal of Applied Social Psychology*, 15(8), 687-699.

Searle, J. R. (1998). The mind and education. In M. Ferrari, & R. J. Sternberg (Eds.), *Self-awareness: Its nature and development* (pp. 3-11). New York: The Guilford Press.

Shahbakhsh, B., & Moallemi, S. (2013). Spiritual intelligence, resiliency, and withdrawal time in clients of methadone maintenance treatment. *International Journal of High Risk Behaviors & Addiction*, 2(3), 132-135. doi: 10.5812/ijhrba.11308

Shaughnessy, J. J., Zechmeister, E. B., & Zechmeister, J. S. (2003). *Research methods in psychology*. New York: Mc Graw-Hill.

Steger, M. F., Frazier, P., Oishi, S., & Kaler, M. (2006). The meaning in life questionnaire: Assessing the presence of and search for meaning in life. *Journal of Counseling Psychology,* 53(1), 80-93. doi: 10.1037/0022-0167.53.1.80

Syed, M., & Juang, L. P. (2014). Ethnic identity, identity coherence, and psychological functioning: Testing basic assumptions of the developmental model. *Cultural Diversity and Ethnic Minority Psychology*, 20(2), 176-190. doi: 10.1037/a0035330

Syed, M., Walker, L. H. M., Lee, R. M., Umaña-Taylor, A. J., Zamboanga, B. L., Schwartz, S. J., . . . Huynh, Q. L (2013). A two-factor model of ethnic identity exploration: Implications for identity coherence and well-being. *Cultural Diversity and Ethnic Minority Psychology*, 19(2), 143-154. doi: 10.1037/a0030564

Templin, J. (2012). *Notes for a lecture on measurement invariance: An introduction to structural equation modeling*. Retrieved from www.jonathantremplin.com

Theron, L. C., Theron, A. M. C., & Malindi, M. J. (2013). Toward an African definition of resilience: A rural South African community's view of resilient Basotho youth. *Journal of Black Psychology*, 39(1), 63-87. doi: 10.1177/0095798412454675

Tirri, K., & Nokelainen, P. (2011). *Measuring Multiple Intelligences and Moral Sensitivities in Education*. Rotterdam: SensePublishers.

Tummala-Narra, P. (2015). Ethnic identity, perceived support, and depressive symptoms among racial minority immigrant-origin adolescents. *American Journal of Orthopsychiatry*, 85(1), 23-33. doi: 10.1037/ort0000022

Umaña-Taylor, A. J., O'Donnell, M., Knight, G. P., Roosa, M. W., Berkel, C., & Nair, R. (2014). Mexican-origin early adolescents' ethnic socialization, ethnic identity, and psychosocial functioning. *The Counseling Psychologist*, 42(2), 170-200. doi: 10.1177/0011000013477903

Umaña-Taylor, A. J. (2015). Ethnic identity research: How far have we come? In C. E. Santos, & A. J. Umaña-Taylor (Eds.), *Studying ethnic identity: Methodological and conceptual approaches across disciplines* (pp.11-26). Washington, DC: American Psychological Association. doi: 10.1037/14618-002

Vaughan, F. (2002). What is spiritual intelligence? *Journal of Humanistic Psychology, 42*(2), 16-33.

Viljoen, H. (2003). Eastern perspectives. In W. Meyer, C. Moore, & H. Viljoen (Eds.), *Personology: from individual to ecosystem* (pp.500-526). Sandown: Heinemann.

Wigglesworth, C. (2006). Why spiritual intelligence is essential to mature leadership? *Conscious Pursuits Inc,* 1-17. Retrieved from http://www.deepchange.com/dig_deeper/index

Wilber, K. (1993). *The spectrum of consciousness*. Wheaton: Quest Books.

Wolman, R. (2001). *Thinking with your soul: Spiritual Intelligence and why it matters*. New York: Harmony Books.

Young, J. S., Cashwell, C. S., & Woolington, V. J. (1998). The Relationship of Spirituality to Cognitive and Moral Development and Purpose in Life: An Exploratory Investigation. *Counseling and Values, 43*(1), 63-69. doi:10.1002/j.2161-007x.1998.tb00961.x

Zohar, D., & Marshall, I. (2001). *Spiritual intelligence: The Ultimate intelligence*. London: Bloomsbury.

Zohar, D., & Marshall, I. (2005). *Spiritual capital: Wealth we can live by*. London: Bloomsbury.

Appendix A

Frequencies and percentages of responses on demographic items in Sample One (N = 1177) and Sample Two (N = 303).

Variables	SAMPLE 1 (*N* = 1177)		SAMPLE 2 (*N* = 303)	
	Frequency	*Percentages (%)*	*Frequency*	*Percentages (%)*
Gender:				
Male	495	47.7	149	49.2
Female	543	52.3	154	50.8
Age:				
18 – 29 years old	498	48	112	37
30 – 39 years old	288	27.8	87	28.7
40 – 49 years old	150	14.5	66	21.8
50 – 59 years old	72	6.9	27	8.9
Above 60 years old	29	2.8	11	3.6
Marital status:				
Single	385	37.1	107	35.3
Committed	108	10.4	21	6.9
Married	513	49.5	170	56.1
Divorced	21	2.0	5	1.7
Separated	7	0.7	-	-
Others	3	0.3	-	-
Education level:				
Certificate of Primary	1	0.1	-	-
Below School Certificate	56	5.4	13	4.3
School Certificate	172	16.6	46	15.2
Higher School Certificate	357	34.5	147	48.5
Prevocational Courses	8	0.7	-	-
Short Courses	13	1.3	-	-
Diploma	118	11.4	38	12.5
Bachelor Degree	196	18.9	34	11.2
Post Graduate Diploma	22	2.1	12	4
Master Degree	80	7.7	12	4
PhD	11	1.1	1	0.3
Others	2	0.2	-	-

	SAMPLE 1 (*N* = 1177)		SAMPLE 2 (*N* = 303)	
Variables	*Frequency*	*Percentages (%)*	*Frequency*	*Percentages (%)*
Occupational background:				
1. Director/Senior Management	21	2	-	-
2. Middle management/ Administrative	68	6.6	-	-
3. Executive Professional (Doctor/Lawyer/architect	17	1.6	-	-
etc.)	121	11.7	-	-
4. Secretarial / clerical	179	17.3	26	8.6
5. Self-employed/	20	1.9	-	-
businessman	127	12.2	104	34.4
6. Artist/ writer/ musician				
7. Teacher (Primary and secondary)	70	6.8	25	8.3
8. Technical (skilled and	23	2.2	-	-
manual)	8	0.8	59	19.5
9. Lecturer/ Professor	16	1.5	34	11.3
10. Nursing sector	6	0.6	-	-
11. Tourism sector	4	0.4	-	-
12. Accounting sector	13	1.2	-	-
13. Public officer	5	0.5	-	-
14. Employed in Private	7	0.7	-	-
sector	5	0.5	-	-
15. Consultant	52	5	-	-
16. Social worker	16	1.5	-	-
17. Unskilled worker	107	10.4	42	13.9
18. Housewife	71	6.8	12	4
19. Retired	81	7.8	-	-
20. Full-time Student				
21. Unemployed				
22. Others				
Religion of participants:	368	35.5	105	34.7
Christianity	64	6.2	3	1
Buddhism	270	26.1	122	40.3
Hinduism	316	30.5	71	23.4
Islam	18	1.7	2	0.7
Others				

Ethnicity:				
Hindu-Mauritian	330	31.8	125	41.3
Muslim-Mauritian	308	29.7	70	23.1
Creole-Mauritian	338	32.6	99	32.7
Mixed background	27	2.6	-	-
Others	34	3.3	9	3

Appendix B

1. THE MULTICULTURAL SPIRITUAL INTELLIGENCE SCALE I (MSIS-1)

S/N	
1	I have a sense of an inner or spiritual life
2	At the end of the day, I tend to reflect on the events that occurred
3	I have a strong feeling of the presence of a Higher Consciousness[4]
4	I am at ease with silence
5	I can face uncomfortable truths about myself.
6	I am aware of my life purpose
7	I am little aware of the values most important to me
8	I do not pay attention to my own thoughts
9	I rarely listen to my little voice that guides me
10	I am aware of my weaknesses
11	I draw on my strength to help others
12	I recognize the voice of the Higher Self speaking to me
13	I am aware when a malignant or non-malignant spirit takes possession of me
14	Inner quietness is important for me
15	I am aware of my true nature
16	I can recognize the voice of my ancestors addressing me
17	I always follow my intuition or gut feeling
18	I allow myself to be open and vulnerable with others
19	I take time to play with young children
20	I feel the presence of a Higher Consciousness very real
21	I am aware of the direction in which a Higher Consciousness is guiding me
22	I consider my dreams useful in gaining insight/clear perception of my life

---

[4] Higher Consciousness/Higher Being/Higher Self/Higher Power/God as you defined it.

23	I listen deeply to the unspoken words in a conversation
24	I do not pay attention to what others have to say
25	I remember to consider what is hidden in a discussion
26	I am able to see things from the other person's perspectives
27	I live and act with consciousness of my mortality
28	In meetings or gatherings, I tend to pause several times to reassess the situation
29	I frequently monitor and notice my thoughts and emotions
30	In times of conflict, I tend to look at a common ground
31	I can hold as true and integrated contradictory points of view.
32	It is difficult for me to enter higher states of consciousness[5]
33	I move freely from different states of awareness
34	I decide when to enter different states of consciousness
35	Higher states of awareness bring clarity to issues
36	I am highly aware of a profound connection between me and other people
37	I tend to look for the relationship among different things
38	I strive/try hard to look at problems from a wide perspective
39	I feel that issues coming in my way are all interconnected
40	I have experiences of knowing the unspoken thoughts of others
41	I am aware that all life is interconnected
42	I experienced transcendental oneness with all living things (Transcendental- to go beyond the material or physical)
43	I am highly aware that a Higher Consciousness is guiding me
44	My faith in life helps me meet daily challenges
45	I am aware of a Higher Self speaking to me for decision making
46	My daily spiritual practice helps me to address stressful situations
47	Spiritual places help me to align with the sacred
48	I am committed to spending time in spiritual practices

---

[5] Higher states of consciousness or feeling of oneness/unity that is felt in for example meditation or other spiritual practices

49	I am aware when I am in the presence of a malignant or non-malignant Spirit
50	I am able to manifest the Higher Self in my physical body
51	I prefer natural healing methods[6]
52	Natural healing method is not at all helpful compared to medication in times of personal difficulties
53	Rites and rituals help me connect with the nonmaterial world
54	Objects that remind of the spiritual help me feel centred
55	Ceremonies and rituals are part of my spiritual practice when something important happens to me
56	Prayer helps me cope with personal difficulties
57	I usually empathise with others (Emphasise-feel compassion)
58	I have an overwhelming experience of nurturing love
59	I cannot emphasise with the pain of someone who disagrees with me
60	I have experienced a sense of reverence/respect for all living beings
61	I am able to feel the joy of others
62	I am able to sit with the pain of others
63	I am indifferent to others' feelings
64	It is difficult for me to acknowledge that I made a mistake
65	I am open to the suggestion of others
66	I recognize that I have limitations
67	I have the strong impression that a higher power is guiding me
68	I am able to admit my mistakes gracefully
69	It is difficult for me to feel compassion for people I meet
70	My ability to connect with others help me to be more effective
71	I am committed to work towards the increased in others awareness
72	My connection with others helps me to be more compassionate
73	In times of stress, aligning my actions with my values are helpful
74	I tend to blame others for my failures

---

[6] Natural healing methods for example healing by prayer/meditation/yoga etc.

75	My ability to find meaning and purpose in life helps me adapt to adverse situations
76	I tend to learn from my mistakes
77	It is difficult for me to bounce back after a failure
78	Despite pressures from outside, I still cling to my ideals
79	My faith tends to quiver after a period of uncertainties
80	In periods of darkness/despair, I draw on my energy from my spiritual values
81	Attending to the nonmaterial aspects of my life remains important despite failures
82	I tend to lose faith quickly in hard times
83	Relying on religious or spiritual leaders to overcome adverse situations is important for me
84	My faith in the higher consciousness strengthens after hard times
85	Attending to spiritual ceremonies is helpful in times personal difficulties
86	My ability to bounce back is strengthened by my spiritual practices
87	My connection with the Higher Self helps me to see clearly in hard times
88	My spiritual practices help me to recover after tough times
89	I have a clear purpose in life
90	My life has no meaning
91	I am seeking meaning in my life
92	I tend to find meaning in everything I do
93	I am able to find every opportunity as a blessing
94	My life purpose is still unclear
95	I feel that each of us contributes to something greater
96	I am aware that my life has a sense of direction that I should follow
97	I feel a sense of vocation to help others
98	I feel grateful for the gifts of life
99	I have spent time reflecting on how I could contribute to my community in a meaningful way
100	I am unlikely to commit myself for the good of others

101	Finding meaning in situations helps me to cope with it.
102	Looking for meanings in times of conflict help me to calm down
103	It is important for me to help others
104	I rarely follow my moral values
105	I am inspired by great leader(s)
106	I have often reflected upon the purpose or meaning of my existence

Self-awareness scale: items 1 – 16
Level of consciousness scale: items 17-35
Transcendental awareness scale: items 36 - 56
Sensitivity scale: 57 - 72
Resilience scale: items 73 - 88
Quest for meaning scale: items 89 – 106

2. THE MULTICULTURAL SPIRITUAL INTELLIGENCE SCALE II (MSIS-2)

S/N	Items
1	I am at ease with silence
2	I can face uncomfortable truths about myself.
3	I am aware of my weaknesses
4	I draw on my strength to help others
5	Inner quietness is important for me
6	I take time to play with young children
7	I am able to see things from the other person's perspectives
8	I am highly aware of a profound connection between me and other people
9	I have experiences of knowing the unspoken thoughts of others
10	I experienced transcendental oneness with all living things (Transcendental- to go beyond the material or physical)
11	I am able to manifest the Higher Self in my physical body
12	Objects that remind of the spiritual help me feel centred
13	Ceremonies and rituals are part of my spiritual practice when something important happens to me
14	Prayer helps me cope with personal difficulties
15	I usually empathise with others (Emphasise-feel compassion)
16	I have an overwhelming experience of nurturing love
17	I am able to feel the joy of others
18	I am open to the suggestion of others
19	I recognize that I have limitations
20	My ability to connect with others help me to be more effective
21	I tend to learn from my mistakes
22	Relying on religious or spiritual leaders to overcome adverse situations is important for me
23	My ability to bounce back is strengthened by my spiritual practices
24	I am able to find every opportunity as a blessing
25	I feel a sense of vocation to help others
26	I have spent time reflecting on how I could contribute to my community in a meaningful way
27	Finding meaning in situations helps me to cope with it.
28	Looking for meanings in times of conflict help me to calm down
29	I am inspired by great leader(s)